Diego Moreira Cassalha

Identifying sustainability criteria in buildings

Diego Moreira Cassalha

Identifying sustainability criteria in buildings

Based on environmental certifications

ScienciaScripts

Imprint
Any brand names and product names mentioned in this book are subject to trademark, brand or patent protection and are trademarks or registered trademarks of their respective holders. The use of brand names, product names, common names, trade names, product descriptions etc. even without a particular marking in this work is in no way to be construed to mean that such names may be regarded as unrestricted in respect of trademark and brand protection legislation and could thus be used by anyone.

Cover image: www.ingimage.com

This book is a translation from the original published under ISBN 978-613-9-72564-9.

Publisher:
Sciencia Scripts
is a trademark of
Dodo Books Indian Ocean Ltd. and OmniScriptum S.R.L publishing group

120 High Road, East Finchley, London, N2 9ED, United Kingdom
Str. Armeneasca 28/1, office 1, Chisinau MD-2012, Republic of Moldova, Europe
Printed at: see last page
ISBN: 978-620-7-90495-2

SUMMARY

Increasingly gaining focus and importance in the field of engineering, sustainable buildings are already present in the country's major urban centres. With various technologies and new processes and materials used in their construction, these buildings stand out from other developments that don't have these devices. As a result, various green certifications and seals characterise and certify these buildings, giving them an even greater degree of competitiveness, engineering and even marketing. The main objective of this work was to create sustainable criteria for evaluating any type of building, based on the guidelines indicated by environmental and regulatory bodies. In order to fulfil this objective, three existing green certifications used in Brazil were used as a technical basis: LEED, AQUA and Caixa Econômica Federal's Blue House Seal, as well as current standards and legislation on the subject. Based on this information, a list of different categories was created to evaluate the sustainability criteria. To this end, the three aforementioned certifications were studied separately and a comparison was made between them. Once the criteria had been defined, a questionnaire was applied to teachers and professionals with market experience, asking them to assess the relevance of each one, with the aim of validating them and defining an applicable value for the final assessment of the criteria in practical cases. The results were an analysis and indication of sustainable buildings and other terms such as the life cycle of buildings and the three pillars of sustainability. It was possible to validate all the criteria selected for the work with a relevance acceptable to the interviewees, as well as defining the cut-off point to be used as a definition parameter in the assessment of buildings aimed at the sustainability of the entire executive process. The aim is to apply these criteria in future engineering and architecture projects in order to explore new technologies and techniques for building projects with a low environmental, social and economic impact on cities.

Keywords: Sustainability in buildings. Assessment criteria. Environmental certifications.

SUMMARY

CHAPTER 1

INTRODUCTION

Since the dawn of humanity, it has been known that man has sought to survive and evolve in his environment by perfecting hunting, security and housing techniques, among others that are so important for his survival, and even without knowing it, he has done this with the minimum of damage to the environment. Since then, evolution has been working on man and he has been in continuous search of new technologies and innovations that help him thrive and live in the current conditions of the planet. One of the most important fields of progress is civil engineering, from which new systems, materials, diverse construction techniques and other inventions that are very advanced even for today's times are constantly emerging.

Nowadays, construction is seen as one of the key sectors in a country's economy, generating countless direct and indirect jobs, increasing the income of both public and private companies and being a laboratory for the development of new technologies. On the other hand, civil construction is also seen as the sector that generates the most waste, from renovations, new constructions and public works, and as one of the planet's biggest consumers of natural resources, causing major environmental impacts. In the 1950s and 1960s, ideas for global sustainability began to emerge in an attempt to reverse this reality. In mid-1972, at the Stockholm Conference in Sweden, the theme of environmental awareness emerged, with the aim of minimising damage to the environment, such as reducing polluting gases in the atmosphere, reducing excessive consumption of natural resources, especially water, and encouraging society to create viable alternatives for a more sustainable and environmentally friendly world.

With these criteria in mind, various construction modifications have been and are being developed to minimise the consumption of water and energy resources and reduce the continuous use of materials used in construction, gradually improving sustainability criteria. Within this context, it can be seen that sustainability goes beyond construction issues and also encompasses social and economic issues. Based on this, the three pillars of sustainability emerge, which try to balance all the concepts and regulations so that the world develops equally within the precepts of the economy, society and the environment, without one interfering with the progress of the other, or when this interference does exist, that it is beneficial.

With a concept of sustainability already formed by today's society, what are the evaluation criteria that can be used to check whether a building is truly sustainable? And what indicators can be used to properly assess these criteria? The aim of this research is to analyse these criteria, compare some of the main existing methodologies for assessing sustainability and analyse the indicators and quality seals most commonly used in the country to categorise buildings as properly sustainable.

CHAPTER 2

RESEARCH GUIDELINES

The guidelines for developing the work are described in the following sections:

2.1 RESEARCH QUESTIONS

The research guidelines of this study are: among the main methodologies of authors specialising in the subject, which criteria can be effectively used to evaluate a sustainable building? Is it possible to create new evaluation methods based on the LEED, AQUA and Casa Azul certifications, in addition to the existing criteria for sustainable buildings?

2.2 RESEARCH OBJECTIVES

The objectives of this research are classified into two types, primary and secondary, and are described below:

2.2.1 Main objective

The main objective of the work is to create sustainability assessment criteria for different types of buildings through mandatory minimum criteria targets.

2.2.2 Secondary objectives

The secondary objectives of the work are:

a) identify the main mandatory criteria required for the creation of a set of criteria and check the most commonly used methodologies for setting up a system of weights and results for these criteria;

b) reference and discuss the meaning of so-called sustainable buildings and what criteria or characteristics should be addressed in order for them to be described in this way;

c) individually analyse and then compare the most widely used sustainability certification assessment methods in Brazil: LEED, AQUA and Caixa Econômica Federal's Casa Azul.

2.3 DELINEATION

The work presented below was carried out in the following stages:

a) Bibliographical research: the research was carried out by reviewing materials by various authors and recent academic articles related to the topic. This stage was carried out throughout the process of preparing the work and served as a basis for all the analyses carried out.

b) Development: the work was based on the ideals and methodologies studied by the various authors referenced. These ideals served as the foundation for proposing a new evaluation method for categorising buildings as sustainable. This stage of the work was developed based on a set of pre-determined criteria and an evaluation questionnaire that sought to measure the evaluative weights of each item specified in the criteria, with the aim of obtaining a response and technical opinion from various civil engineering professionals, such as professors, scholars in the field, entrepreneurs in the construction industry and civil and environmental engineers. With the result of this material, the criteria were assembled in full, with the future possibility of applying them to projects in the design phase and even those already built, judging their sustainable effectiveness;

c) Final considerations: finally, all the considerations relating to the aforementioned objectives of the article have been made.

2.4 DELIMITATION

The criteria defined were only created and measured by choosing the sustainability evaluation criteria adopted through the basic criteria required by international bodies and legislation on building sustainability, as well as the parameters adopted by the LEED, AQUA and Caixa Blue Seal certifications.

2.5 LIMITATION

The limitation of this work is the absence of the actual application stage of the criteria at this point in the work, because they need to be applied from the preliminary design stage until the building is used by users, and this process takes a long time to complete.

2.6 FLOWCHART

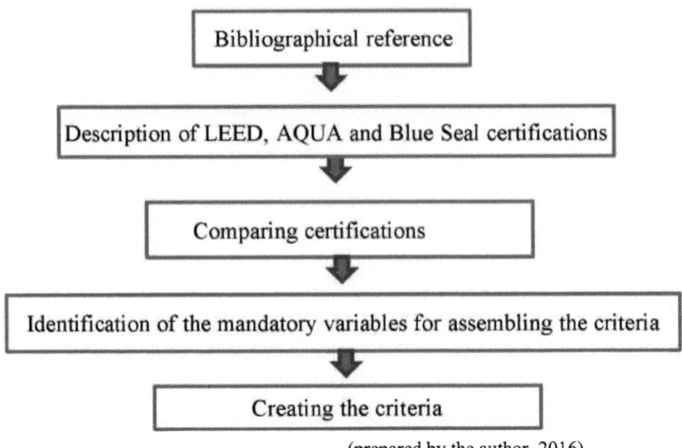

(prepared by the author, 2016)

CHAPTER 3

THEORETICAL FRAMEWORK

Even in prehistoric times, man developed rapidly and changed his surroundings as a result. According to Dias (2011), after the emergence of agriculture and sedentary lifestyle, there was a profound change in the relationship between man and nature, where necessity led individuals to gather in groups in order to maintain their quality of life. As a result of these agglomerations, environmental degradation has arisen, with the gradual disappearance of some species. Lemos (2013) says that from the second half of the Middle Ages, with the growth of the world's population, through the Industrial Revolution, with the agglomeration of people in large urban centres, until the end of World War II, when pollution grew to alarming levels as a result of the fighting, environmental awareness was still just a seed planted in the minds of some scholars. With the development of the economies of the world's major powers since the beginning of the 20th century, driven by the main sectors: industry, agriculture, services and construction, environmental problems began to worsen even more. Disordered consumption of natural resources; unconscious waste of raw materials and energy, such as electricity; pollution of water and air resources; increased generation of waste from industrial and civil processes and many other problems began to become routine and more noticeable to society, generating health, visual, environmental and economic impacts in large urban centres. As a way of reducing this growing damage, new concepts are beginning to be used and implemented by modern society.

Faced with the reality of the time, 1972 saw the first major world conference that openly addressed the impacts caused to the environment. Gouvinhas (2008) describes that the Stockholm Conference took place in Sweden's capital, Stockholm, highlighting these environmental impacts and what aggravating factors could worsen this scenario in the coming years, bringing all countries face to face to debate solutions aimed at reducing damage to the environment in the following years, without having major impacts on their economies and society. Environmental awareness was born, and soon afterwards it would take the first steps towards a sustainable world. According to Leite (2011), the Stockholm Conference put the issue of sustainability definitively on the world political agenda, since environmental phenomena such as global warming and the destruction of the ozone layer had become problems of global significance.

Exactly ten years later, this time in Noiróbi, Kenya, the United Nations Organisation (UNO) held its second conference on the subject, where it revealed that the planet's conditions were more serious than expected and ran counter to the development of the great powers, which were the main villains in the destruction of the planet. Leite (2011) mentions that in 1987, the UN once again held a meeting on the environment at the World Commission on the Environment (UNCED), where one of the most important reports on the subject was produced, called "Our Common Future" or the "*Brundtland Report*". The Report described how it was possible to combine the development of countries without generating major impacts on the environment and on the economy and society, giving rise to the three pillars of sustainability (Figure 1). This figure shows that there is no sustainability unless there is a balance between the three pillars, i.e. sustainability will only exist when none of the pillars stands out or is inferior to the others.

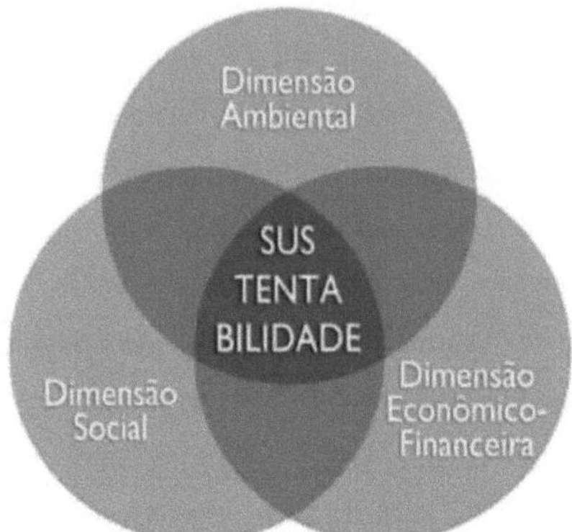

Figure 1 - Three pillars of sustainability

(source: Collective Sustainability, 2013)

Named in honour of the Prime Minister of Norway, Gro Harlem Brundtland, head of the Commission in Noiróbi, the report contained global data on climate warming and the destruction of the ozone layer, hitherto new topics in this scenario. The Brundtland Report (1987) also contained a number of approaches to action to be taken by the European Union.

countries involved to reduce the parameters mentioned above, among others that impacted on the environment. Some of the measures described in the Report (1987) were:

- limiting population growth;
- guaranteeing basic resources for humanity in the long term (water, food, electricity);
- preservation of ecosystems and biodiversity at a global level;
- meeting basic human needs (health, education, housing, security);
- controlling disorderly urbanisation in urban centres, with the aim of combining the countryside and smaller towns;

According to Dias (2011), the document closely links economics and ecology and establishes an axis around which sustainable development should be discussed. With regard to the term sustainable works, the Brundtland

Report (1987) mentions the following items:

- use of new construction materials;
- restructuring urban and industrial areas to improve and make the best use of available space;
- use of alternative sources of energy generation (wind, solar);
- reusing recyclable materials;
- rational water consumption;
- reducing the use of chemical products to reduce damage to workers' health.

Even more recently, in 1992, Eco-92 was organised in Rio de Janeiro, where the theme of sustainability was enshrined and where the United States of America (USA), contrary to the other countries involved, did not want to be part of the implementation of the timetable aimed at reducing carbon dioxide (CO_2) emissions and signing the convention on global biodiversity. Leite (2011) describes that Eco-92 formalised the Earth Charter, the Declaration on Forests, the Convention on Biological Diversity, the Framework Convention on Climate Change and the Convention on Climate Change.

Agenda 21. Each of these documents dealt with specific issues, such as creating new levels of cooperation between developed and developing countries and stabilising the level of greenhouse gas generation. Agenda 21 developed a programme of actions to make the principles established at the meeting feasible and practical. Dias (2011) mentions that in addition to Agenda 21, other documents were developed with the same environmental objectives, such as the Convention on Biological Diversity, the Principles for Sustainable Forest Management and the Rio de Janeiro Declaration on Environment and Development.

Not least in the 1990s, the Kyoto Protocol was also developed. It represented the historic point of redefinition of world growth patterns by incorporating the first official definition, based on scientific data, of the concept of sustainable development (GOUVINHAS, 2008). Later, in 2002, the second edition of Eco-92 took place, known as Rio+10, where the proposals created in Agenda 21 were re-evaluated and humanity's economic and social needs were redefined.

In 2012, again in Rio de Janeiro, the second edition of the international conference on the environment, Rio+20, took place. According to Lemos (2013), this edition debated current sustainability issues and renewed and reaffirmed the participation of leaders from the various participating countries, with the aim of improving the environment and reducing impacts on the planet. An assessment was made based on ECO-92, which had taken place in the same place 20 years earlier, in order to see what had changed since then. Topics such as combating poverty worldwide, the importance of the Green Economy, actions to guarantee global sustainability and international governance in the development of sustainable technologies and actions were discussed. However, the final outcome of the congress was not what was expected. According to Lemos (2013) there were several impasses between the governments of developed and developing countries, frustrating the world's expectations of a more sustainable planet.

3.1 SUSTAINABLE BUILDINGS

Below, we will discuss some basic concepts about sustainability in buildings, as well as other important themes in this area that are important for understanding the objectives of this work.

3.1. 1Definitions

Sustainable buildings can be understood as: "a holistic process that aspires to restore and maintain harmony between the natural and built environments, and to create settlements that affirm human dignity and encourage economic equity." (AGENDA 21, JOHN et al, 2001)

With sustainable engineering in the spotlight, it is becoming more routine to design and build sustainable buildings that aim to meet the standards specified by the main certifying bodies. But beyond certifications, how can a building be sustainable? The Guide to Sustainability in Construction (MINASCON, 2008) lists three basic parameters for a building to be sustainable:

-Project phase - must have quality and efficiency in its design;

-Execution - the work must be legally correct and all those involved aware of their professional responsibilities;

-Post-work - innovations must be continually sought in the field of construction, giving greater prominence to this issue.

In addition to these prerequisites, the Sustainability Guide (MINASCON, 2008) also lists ideals for seeking positive results in the design, execution and maintenance of sustainable buildings. There are also nine basic guidelines cited by the same author: quality of implementation, management of water use, management of energy use, management of materials and reduction of waste generated, pollution prevention, environmental management of the entire process, management of the quality of the interior environment, quality of services and economic performance.

According to the Institute for the Development of Ecological Housing - IDHEA (2014), sustainable construction is characterised by meeting the following criteria: sustainable construction planning, passive use of natural resources, energy efficiency, correct management and saving of water resources, management of waste generated on the building site, air and work environment quality, thermal and acoustic comfort, rational use of materials, and use of environmentally correct products and technologies. Table 1 below lists the main criteria according to each of the aforementioned guides (IDHEA, 2014) and the Construction Sustainability Guide (MINASCON, 2008).

Table 1 - Comparison of criteria

SUSTAINABILITY GUIDE (2008)	IDHEA (2014)
Quality of implementation	Sustainable construction planning
Water use management	Passive utilisation of natural resources
Energy use management	Energy efficiency
Materials management and waste reduction	Correct management and saving of water resources
Pollution prevention	Management of waste generated on site
Environmental management of the entire process	Air quality and the working environment
Indoor environmental quality management	Thermo-acoustic comfort
Service quality and economic performance	Use of environmentally friendly products and technologies

(source: Author, 2016)

This comparison shows that the search for new materials, recycling and reuse of raw materials, innovative project design, alternative energy sources and new technological inventions are essential as sustainability criteria in civil engineering. These, however, are still hampered by some economic, sustainable construction and social obstacles, which directly affect the evolution of systems. There is still little private initiative to seek economic viability inside and outside companies so that the development and execution of sustainable works in Brazil is more attainable. This barrier is created by outdated policies on environmental protection, a lack of laws favouring companies to carry out such works and even a lack of information on the part of society, which is unaware of all the benefits its cities could gain from such changes. According to Dias (2011), environmental policies developed by specific agencies are increasingly proving ineffective, since the environmental problem involves the most diverse sectors of the economy, which causes policies to become isolated and generates a lack of environmental responsibility in other sectors.

3.1. 2Building life cycle

Thinking about the life cycle that a building will have, from its conception to its completion, is also of paramount importance within the sustainable horizon.

> "The time in which a building is constructed represents a small part of its useful life and so building sustainably means ensuring that, in addition to the planning and implementation phases, the occupation, maintenance and demolition phases contribute to generating fewer impacts" (LEITE, 2011).

For Silva (2003), the life cycle can be described as a procedure that aims to anticipate the occurrence of environmental impacts that may be generated in the different stages of a product's life cycle.

Silva (2003) goes on to describe the main applications of the building life cycle:

- -evaluation of the environmental suitability of a given technology, process or product;
- -identifying possibilities for improving an individual process/product, especially with regard to the efficient use of resources and the reduction of emissions;
- -comparison of technological alternatives, different processes or products, but aimed at the same function;
- -generating information for consumers that will serve as the basis for environmental labelling.

With the often disorganised use of urban spaces, this reality has become more visible over the years, showing the use of vacant land with little or no awareness of the future life cycle of the building that will be built on that site. This, over time, can have major impacts in the future, when its useful life comes to an end and it gives way to another new development. To prevent this from happening, care must be taken when choosing the site where the building will be erected, as well as its surroundings, in a dynamic and cultural way.

According to Leite (2011), the main objectives for a complete life cycle can be described according to the building's phases. Starting at the design stage, one should be aware of the possible lifespan of the project and look for future alternatives for its replacement, taking sustainable practices into account. In the construction phase, it's important to look for rational alternatives for water and energy consumption that don't involve

waste, aim to manage the waste generated and even reuse it elsewhere on the site, choose sustainable materials and even carry out social actions with employees in order to train them in socio-environmental issues.

Also, according to Leite (2011), during the occupation and use phase, the phase with the longest useful life, what should be taken into account most is the correct and periodic general maintenance of the building, where sustainable alternatives should also be sought so that its useful life is extended, reflecting the comfort and well-being of users. Finally, the last phase, demolition to make way for another development, must be carried out in an orderly and thoughtful manner, in order to reduce impacts, reduce waste generation, check for suitable disposal sites and think about recycling and reusing it in other works.

Table 2 - Life cycle of buildings

AVERAGE SERVICE LIFE	CONSTRUCTION PROCESS
1 to 3 years	Building design and construction
3 to 5 years	Maintenance and use time
10 to 15 years	Average time of use and partial renovation
30 to 50 years	Long service life and total renovation

80 to 120 years	Lifespan of the structure
Over 150 years	Lifespan of monuments

(source: Soares, Souza and Pereira, 2006)

Chart 2 shows that the lifespan of a building's structure, given the right preventive maintenance, can last for around a century. It can also be seen that such structural maintenance, as well as other various types of maintenance that buildings need, should be carried out within the first three to five years after the project is handed over. Finally, Soares, Souza and Pereira (2006) point out that it is very important to assess the life cycle of buildings, since there are big differences between the life cycle of industrial products, which last around a week or even days, and buildings, which can last for centuries.

3.1.1 Training

Training all those involved in the processes of designing projects, carrying out construction work and even the future users of developments is a viable alternative that is simple to apply, but which can yield excellent results. Encouraging project professionals to plan buildings to consume less energy, avoid wasting water, utilise the natural climate for thermal comfort and seek better interaction with the external environment is a sustainable solution.

"The first thing needed to produce sustainable construction *is for* everyone involved to get on board, making sure that in each of the areas of intervention the required parameters are met to the greatest possible degree" **(LEITE, 2011).** The biggest obstacle to this resource is the training of construction workers, as they generally have low levels of education and low economic and social status. This makes it more difficult for them to interact with the sustainable environment. However, their training is essential and should be carried out frequently so that their sustainable awareness grows, which will be reflected in their homes, families and neighbours. It is important that they understand the benefits of correctly managing the waste generated in construction processes, avoiding the waste of resources and materials and seeking more technological alternatives for some processes.

The work ends with making building users aware that preventive and periodic maintenance avoids a number of problems that impact the environment, resulting in a reduction in their comfort and well-being over the years.

3.2 SUSTAINABILITY CERTIFICATIONS AND CRITERIA

Evaluation criteria are understood to be the clear indication of certain aspects with the aim of organising, defining and measuring them according to a single system in order to obtain a measurable value of a given subject. "Most sustainability assessment methodologies are based on analysing indicators that cover the various topics considered relevant" (MATEUS, 2004). According to Deponti et al. (2002), an indicator shows, reveals, proposes, suggests, exposes, mentions, advises and reminds us of something. As such, an indicator can be used to measure changes in a particular system, but measurement cannot identify whether this system is growing, stagnating or declining.

Evaluation criteria can be defined by credits that generate evaluation indexes divided into categories that analyse the efficiency of projects internally and in their surroundings, generating classifications with different results. This is the case with LEED (Leadership in Energy and Environmental Design) and BREEAM (Building Research Establishment Environmental Assessment Method). They can also be based on the performance of executive management and the processes employed, giving the development a sustainable or non-sustainable rating, as in the case of the Haute Qualité Environnementale (HQE) certification and the High Environmental Quality Process (AQUA).

Some authors, such as Mateus (2004), Fossati (2008), Deponti (2002) and Silva (2006), have developed work on sustainability assessments in buildings, as well as their own methodologies for analysing the numerous criteria for defining sustainable buildings around the world. Their initial bases are, more often than not, sustainable certifications. Among these international certifications, the most important and widely used are BREEAM, LEED, HQE, the Green Building Challenge (GBC) and the national method, AQUA. According to Mateus (2004), the European Commission was set up in Europe to standardise the certifications that were emerging in the 1980s and 1990s. It defined ten basic indicators for the creation of new sustainability assessment methodologies. These indicators were divided into two categories, main and secondary, and are listed in Table 3.

Table 3 - Sustainability indicators

MAIN INDICATORS	SECONDARY INDICATORS
User satisfaction	Distances to teaching spaces
Impacts on climate change	Sustainable development coordination systems
Mobility and public transport	Noise
Access to service areas and green spaces	Sustainable land use
Air quality	Products that respect sustainable development

(source: Mateus, 2014)

Through these prerequisites, it was possible to draw up some of the aforementioned certifications, and with their help, it was possible to create general criteria for evaluating sustainable buildings. The LEED, BREEAM and GBC methods have a global sustainable analysis theme and require prior knowledge of a range of parameters, most of which are complex to obtain. For this reason, the inclusion of these international certifications in Brazilian construction projects can be complicated, and it is not easy to obtain a coherent and satisfactory final assessed value, mainly because they are based on technologies and methods from developed and more advanced countries, which makes them somewhat unfeasible to solve with the current construction techniques used in the country. According to Leite (2011), other important factors are the economy and society, which have a direct impact on sustainability. Among the certifications, the most widely used is AQUA, a national certification based on the French HQE, which uses a methodology more suited to the reality of Brazilian construction work.

3.2.1 LEED certification

Created in 1994 in the USA and developed by the U.S. Green Building Council (USGBC), a performance rating system was created with the aim of evaluating sustainable construction methods. Based on the example of other countries with similar certification models, such as BREEAM in the UK and BEPAC in Canada, LEED was created with the aim of creating environmentally friendly and profitable developments.

According to the Green Building Council Brasil - GBCB (2001), certification was created to encourage the transformation of building design, construction and operation, always focussing on sustainability. The system is based on a voluntary membership programme and aims to assess the environmental performance of a building considering its life cycle and can be used in any type of development. It is certified at different assessment levels by analysing the documents required to comply with the mandatory items contained therein. These levels are divided into four systems (GBCB, 2001): Basic Certification (from 40 to 49 points), Silver Certification (50 to 59 points), Gold Certification (60 to 79 points) and finally Platinum Certification (80 to 110 points), as shown in Figure 2.

Figure 2 - LEED rating labels

(source: GBCB, 2001)

Since its creation in 1994, LEED has had different versions according to the type of building to be certified, as described in Table 4. These versions define the type of building to be certified, such as small homes, commercial and residential buildings of all types, schools and educational institutions, hospitals, health centres and other healthcare facilities, and also a specific version aimed at urban development, such as planned neighbourhoods. An example of this type of development can be seen in England's capital, London. The

BedZED eco-neighbourhood[1] **(Beddington Zero Energy Development)** is a village of around one hundred houses and has been developed **to use energy efficiency techniques to create a "zero emission community".**

Table 4 - LEED versions

VERSIONS	TYPE OF BUILDING
LEED NC *(New Constructions and Major Renovations)*	New commercial buildings and renovation projects of some scale
LEED H *(Home)*	**Green" housing**
LEED EB *(Existing Buildings)*	Support the sustainable operation and maintenance (and improvement) of existing buildings
LEED Cl *(Commercial Interiors)*	Interior commercial spaces
LEED S *(Schools)*	Aimed at the specific needs of schools
LEED HC *{Healthcare')*	Healthcare spaces
LEED R *(RetaiT)*	Commercial spaces
LEED CS *(Core and Shell Development)*	It covers the construction of building elements such as the structure, envelope and systems
LEED ND *(Neighbourhood Development)*	Aimed at surrounding urban development, based on the concept and principles of *smart growth* (under development)

(source: Lucas, 2011)

The certification works on the basis of the scores achieved at the different assessment levels. According to the GBCB (2001), certification has 7 dimensions to be assessed, all of which have prerequisites and credits. As there is no weighting between the scores, the same assessed building can have a maximum score in one area and a minimum score in another, without affecting its final assessment result. For Lucas (2011), the objective assessment areas are described below:

-Sustainable location : control erosion and reduce negative impacts on water and air quality. Adopt a sedimentation and erosion control plan for the project site during construction;

water efficiency : reducing water consumption by developing efficient irrigation and reuse systems, as well as a water use re-education programme;

-energy and atmosphere: checking and ensuring the essential elements of buildings and that systems are designed, installed and calibrated to operate objectively;

-materials and resources: facilitate the reduction of waste generated by the building's occupants;

[1] BedZED. Source: Sustainable Cities. Available at: http://www.cidadessustentaveis.org.br/boas- praticas/bedzed-liderando-c caminho-no-design-de-eco-bairros. Accessed on: 26 September 2016.

-indoor environmental quality: establishing a minimum indoor air quality performance to prevent the development of problems in buildings arising from indoor environmental quality, while maintaining the health and well-being of occupants;

-innovation and the design process: the use of the aforementioned criteria should not be an obstacle to the designer's creation;

-regional priority: determines the different environmental priorities between different regions.

According to the GBCB (2001), LEED certification is evaluated over a period of five years from the time the building is registered on the committee's website. After this period, it is revalidated by the US organisation USGBC. The system is widely used in the USA, its country of origin, mainly because its assessment criteria are based on various American regulatory and control bodies. Some of the bodies that created the LEED certification criteria were the American Society of Heating Refrigeration and Air Conditioning Engineers (ASHRAE), which controls refrigeration, air conditioning and heating standards; the American Society for Testing and Materials (ASTM), which is the national body for the National Institute of Metrology, Quality and Technology (INMETRO); as well as the U.S. Environmental Protection Agency (EPA) and the U.S. Department of Energy (DOE).

The certification is divided into key areas, which are then subdivided into specific items, as shown in Table 5 below.

Chart 5 - Key areas of the LEED label

KEY AREAS	**CRITERIA**
Site sustainability	Erosion and sedimentation control, site selection, urban redevelopment, networks involving environmentally contaminated sites, transport, reduction of disturbances caused by construction, management of bad weather situations, recovery and protection of open spaces, landscape and exterior design and reduction of direct light radiation output.
Water management	Efficient use of water, innovative treatment technologies.
Energy and atmosphere	Fundamental instructions for building systems, minimum energy performance, reduction of **CFCs, renewable energies, additional instructions,**
	measurement and verification, green energy and ozone depletion.

Materials and resources	Collection and storage of recyclable materials, reuse of the building, construction waste management, reuse of resources, recycled content of materials, local/regional materials, materials quickly and certified wood.
Internal environmental quality	Information on innovative measures incorporated into the project and their sustainable benefits.
Innovation and design processes	Minimum indoor air quality performance, indoor tobacco smoke control, carbon dioxide monitoring, increased ventilation efficiency, indoor air quality management plan, **low-VOC** materials, **ability to control** systems, thermal comfort, natural lighting and views.

(source: Lucas, 2011)

In order to receive certification, the steps mentioned by the GBCB (2001) must be followed, in which the developer must fill in a registration form containing all the information about the building to be analysed and the company responsible for its execution, which can be found online on the official website. Based on this information, the possibility of sustainable certification for the development is checked. Subsequently, all the required documentation is requested from the company responsible so that the indicators can be analysed. Once all the documentation is correct and the work is 100 per cent complete, the final certification inspection is carried out. These processes take around three months for the design phase and three to six months for the final inspection of the building. The costs for this certification average between 5 and 10 per cent of the total value of the building and are described in Table 1.

Table 1 - LEED implementation costs

LEED IMPLEMENTATION COSTS	
Registration fee for evaluation U$D600.00	
Additives	
Projects of up to 5,000 m^2	U$D 2,250.00
From 5,000 to 50,000 m2	U$D 0.45 per itf
Over 50 thousand m2	U$D22,500.00
Consultancy	
Approximately 1% of the total cost of the work	

(source: Lucas, 2011)

One of the possible reasons why LEED certification is not so common in Brazil is its high cost for large property developments (over 50,000 m^2 of built area). Another setback in choosing LEED is that its values are calculated in US dollars, since its variation according to the world economy can make choosing this certification even more expensive.

In Brazil, the choice for this certification has been gradually growing in various developments located mainly in the southeast and centre-west regions of the country. Data from the GBCB (2014) shows the growing demand for certification, as shown in Figure 3 below, which shows the demand for the main LEED certified categories.

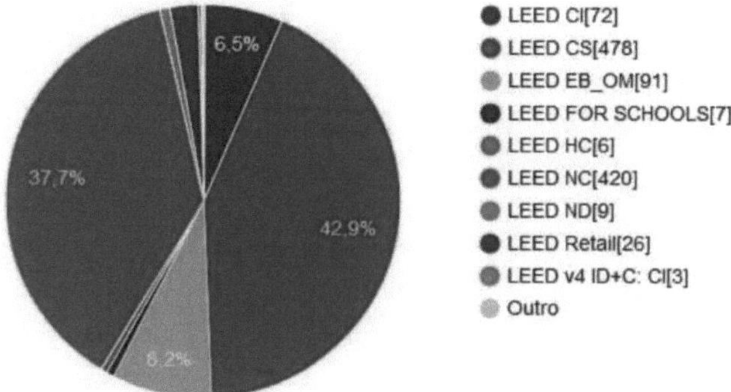

Figure 3: Registrations by LEED category
(source: GBCB, 2014)

Figure 3 shows that demand for LEED certification is highest in the LEED CS subdivisions, i.e. the construction of different elements of buildings, such as the structure, envelope and internal systems such as energy, ventilation and water. This is followed by LEED NC, which represents new commercial construction or various renovations. There is also a small, almost insignificant demand for certification in governmental projects, such as LEED S, for schools, LEED ND, which aims to develop urban areas as a whole, and LEED HC, for public and private healthcare facilities.

With regard to the growing number of LEED certification registrations in buildings across the country, there has been a relatively high demand over the years. This demand began in 2010 and continues until the present year. Another positive point was the increase in the number of certificates accumulated over the same period mentioned above. Figure 4 shows the evolution of the number of LEED certification registrations in Brazil with a projection to the end of 2016.

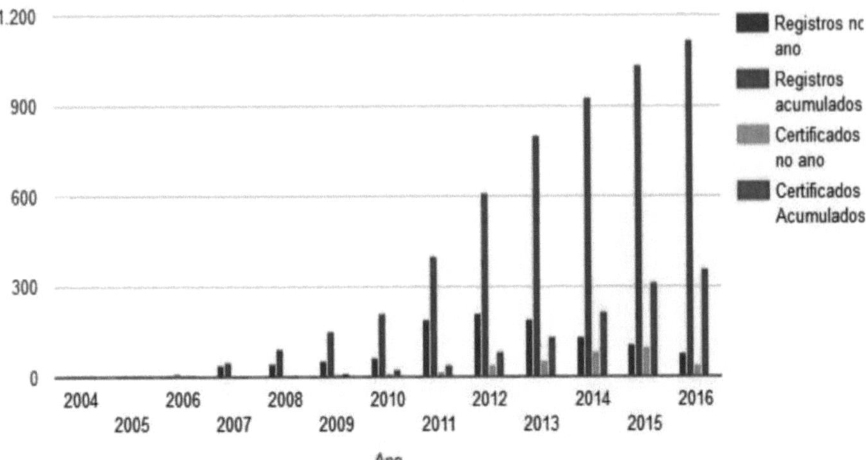

Figure 4: LEED registrations in Brazil

(source: GBCB, 2014)

Figure 4 shows the growing demand for certification over the last six years (from 2010 to 2016), which indicates the search for improvements in the construction sector and the qualification of new buildings, which has been gaining ground with companies' concern for sustainability.

Finally, Figure 5 describes the places where LEED certification is most widespread, with the states on the Rio - São Paulo axis being the biggest consumers. This confirms the location of the greatest concentration of wealth and economic development in the country. Data from Exame magazine (2012) shows that around 65% of Brazil's Gross Domestic Product (GDP) is concentrated in five states: São Paulo, Rio de Janeiro, Minas Gerais, Rio Grande do Sul and Paraná.

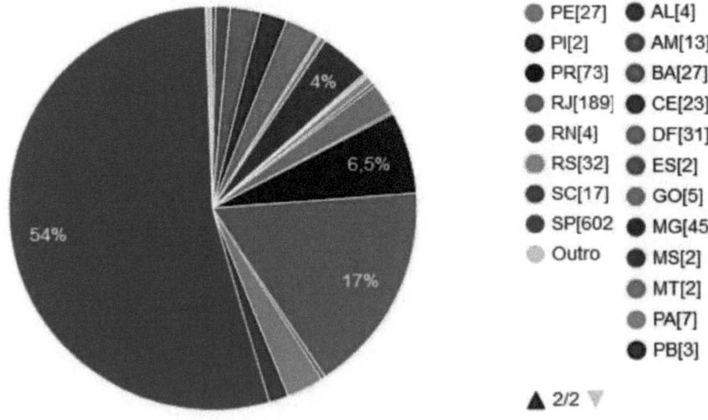

Figure 5: LEED registrations by state

(source: GBCB, 2014)

3.2.2 AQUA certification

The Brazilian AQUA certification, based on the French HQE, was developed in 2008 by the Carlos Alberto Vanzolini Foundation in partnership with the Polytechnic School of the University of São Paulo and the Centre Scientifique et Technique du Bâtiment (CSTB). The certification process aims to guarantee the environmental quality of a new construction or refurbishment project using independent audits (LEITE, 2011).

According to the Vanzolini Foundation (2015), the AQUA process proposes a new approach to building sustainability in Brazil, seeking technical development in the areas of culture, climate, technical standards and current regulations. What sets this seal apart is the intention to assess the efficiency of the measures adopted at different stages of the project - programme, design, implementation and operation. This assessment is measured qualitatively, not quantitatively, and not only at the design stage, but also audits are carried out at the construction stage to assess the choices made (UCHOA, 2014).

This means that, unlike other certifications such as LEED, BREEAM and HQE, the assessment is estimated in quality levels and not in numerical values. The levels are described below (UCHOA, 2014):

-Excellent : maximum performance seen in projects;

-Superior : good sustainability practices;

-Good : minimum acceptable performance for a sustainable enterprise.

AQUA certification aims to benefit both the developer, socio-environmental bodies and the buyer of the property, as can be seen in Chart 6.

Chart 6 - Benefits of the AQUA label

AQUA CERTIFICATION BENEFITS	
Entrepreneur	Proves the high environmental quality of the building Differentiates your portfolio in the market Increases the speed of sales or rentals Maintains the long-term value of the property Associates *the* company's image with AQUA Improves relations with environmental organisations and the community
Buyer	Direct savings on water and electricity Lower condominium fees (water, energy, maintenance and upkeep)

	Improved comfort, health and aesthetic conditions Greater property value over time
Social and environmental organisations	Lower energy and water consumption Reducing greenhouse gas emissions Reducing pollution Reducing waste production

(source: Leite, 2011)

According to the Vanzolini Foundation (2015), the AQUA label is assessed by two individual systems that target specific items: the SGE (Development Management System), which assesses the management system implemented, and the QAE (Environmental Quality of the Building), which assesses the architectural and technical performance of the development. The certification process is divided into stages, which are described in the following chronological order according to the Vanzolini Foundation (2015):

- Pre-project phase: site analysis; prioritisation of the 14 categories; justification and proposal for the EQ profile; planning of the SGE and evaluation of the EQ;
- Project phase: drawing up project solutions; managing the project according to the SGE and evaluating the QAE;
- Execution phase: execution of the work in accordance with the SGE; management of material control records and impacts generated on site; training of users and managers of the future development; commissioning and final evaluation of the QAE.

With this, the developer will receive two certificates: one from the Vanzolini Foundation's AQUA process and the other from Cerway HQE. It is important to emphasise that once the building has been completed, the managers are responsible for maintaining the seals in accordance with the requirements of the assessment body. According to the Vanzolini Foundation (2015), some good practices should be adopted throughout the building's useful life, such as: periodic maintenance of ventilation, heating and air conditioning systems; preventive maintenance of water and sewage networks, as well as their management; maintenance and management of electrical systems, lifts, automatic garage doors and security systems; cleaning and maintenance of communal and external areas; and finally, signposting and guidance on specific areas for sorting the waste generated.

Tables 2 and 3, according to the Vanzolini Foundation (2015), show the total number of certified projects over the years, as well as the main states where certification is most common.

Table 2: Total AQUA certifications per year

Year	Total Developments
2009	5
2010	20
2011	36

2012		71
2013		157
2014		203
2015		231
2016		235

(source: Vanzolini Foundation, 2015)

Table 3: States with the widest range of AQUA certifications

States	ENDEAVOURS
AL	1
AM	1
BA	5
EC	5
DF	7
ES	2
MG	5
MS	1
PB	3
PE	2
PR	9
RJ	30
RS	4
SC	2
SP	156
Grand total	**235**

(source: Vanzolini Foundation, 2015)

Table 3 shows that, like LEED certification, the AQUA seal also concentrates its demand in the states of São Paulo and Rio de Janeiro, reinforcing the idea adopted earlier about the concentration of wealth and development in the country.

According to the Vanzolini Foundation (2015), implementation costs range from the pre-project phase to the final execution of the project. These costs include face-to-face audits, assessments and the issuing of certifications, and vary in value from R$17,500.00 for built-up areas of up to 1,500m^2 , to R$87,500.00 for built-up areas of 4,500m2 or more.

"One of the limitations of AQUA is that it only has technical references for residential buildings, neighbourhoods, allotments and service sector buildings such as offices and schools." (UCHOA, 2014). From a practical point of view, this may not be an inconvenience, since it is possible to apply or even create new methodologies for developments that do not fall within the scope of the certification and thus use these parameters for buildings of the same model that will emerge in the future, adding them to the current AQUA model.

3.2.3 Blue House certification

The Blue House label was created by the Caixa Económica Federal - CEF bank to assess works carried out through its financing, as mentioned in the Caixa Guide - Environmental Sustainability (2010). The aim of its creation was to guarantee sustainable development and the well-being of society. According to Almeida, Viana and Pisoni (2014), as well as encouraging good practices in terms of sustainability in the construction industry, the label helps to promote a sustainable environment through public or private works, including environmental protection at a national level, social justice and economic viability.

> "The certification analyses each project, checking for aspects linked to environmental sustainability in the projects to be financed by the bank, such as new construction systems (building more using fewer materials, using waste instead of natural raw materials, better adapting the built environment to its surroundings); new systems for using water (drinking water and waste water); as well as new energy generation systems" (ALMEIDA; VIANA, 2014, p. 3).

There are six different categories for classifying buildings. According to the Caixa Guide - Environmental Sustainability (2010), these are described below: Urban quality (5 criteria); design and comfort (11 criteria); energy efficiency (8 criteria); conservation of material resources (10 criteria); water management (8 criteria); social practices (11 criteria). Certification can be awarded to the development in three different categories, according to the criteria (compulsory or not) achieved: Bronze Seal (19 compulsory criteria); Silver Seal (19 compulsory plus 6 free); Gold Seal (12 free criteria in addition to the 19 compulsory).

Chart 7 shows the mandatory criteria for acquiring the bronze certification of the Casa Azul Seal, i.e. the most basic certification that must be achieved in projects financed by the bank, according to the Caixa Guide - Environmental Sustainability (2010).

Table 7 - Mandatory criteria

CATEGORY	MANDATORY CRITERIA
Urban quality	Quality of surroundings - infrastructure Quality of surroundings - impacts
Design and comfort	Landscaping Place for selective collection Leisure, social and sports facilities Thermal performance - seals Thermal performance - sun and wind orientation
Energy efficiency	Energy-saving devices - common areas Individualised metering - gas
Conservation of material resources	Quality of materials and components Reused moulds and props Construction and demolition waste management
Water management	Individualised metering - water Economising devices - flushing system Permeable areas
Social practices	Education for CDW management

	Environmental education for employees Orientation for residents

(source: Guia Caixa, 2010)

According to the Caixa Guide (2010), an important factor in certification is the correct preparation of the owner's manual, which must contain important information about the Casa Azul Seal criteria. Some of these criteria are described as the correct use and maintenance of the systems and equipment installed, leaving the administrator and/or property manager primarily responsible for maintaining and preserving the certification over the years, since the seal can be lost if there are future disagreements with the criteria used to obtain it.

According to the Caixa Guide (2010), one of the main objectives of the Blue House Label is to create communities that meet the needs of their residents. To this end, the Guide is based on the system called Disco Egan, which was a review of the various skills needed to implement the first sustainability programmes in England. The disc presents eight essential components for sustainability in communities: governance; connectivity; availability of services; environmental responsibility; justice/equality; prosperity; design and construction; and livability, inclusion and security, as can be seen in Figure 6.

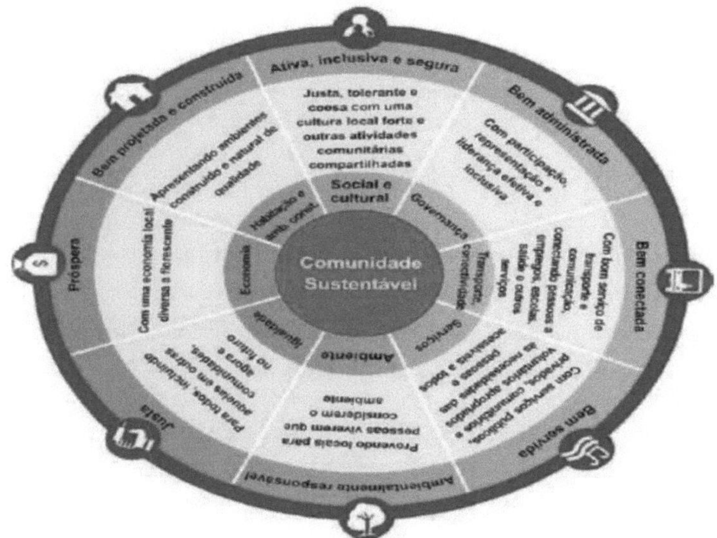

Figure 6: Egan's disc

(source: Egan 2004 **apud** Guia Caixa (2010))

According to the Caixa Guide (2010), four items on the disc (governance, justice/equality, prosperity and livability, inclusion and security) are indirectly linked to the interventions of designers, developers and construction companies. The other items (connectivity, availability of services, environmental responsibility and design and construction) are focused on the planning and production of new housing developments. As such, the Casa Azul Seal aims to contribute to the development of more sustainable developments, and the criteria that the certification requires are based on this assumption.In short, as the Guide mentions, each criterion has its own main objective, shown in Table 8.

Chart 8: Socio-economic criteria of the Casa Azul Seal

CRITERION	OBJECTIVE
Quality of surroundings - infrastructure	To provide quality of life, services, community facilities and commerce in the vicinity of the buildings; To seek the well-being, health and safety of the residents near the development.
Improvements to the surroundings	Encouraging aesthetic, functional, landscaping and accessibility improvements for the population.
Recovering degraded areas	Encouraging the recovery of degraded social/environmental areas.
Property rehabilitation	Encouraging the occupation of urban voids in order to return disused buildings to the environment, the economic cycle and urban dynamics.

(source: Guia Caixa, 2010)

3.2.4 Comparison of certifications

Of the three most widely used certifications in the country, as presented above, a comparison can be made by identifying the criteria adopted in each certification in the following areas: cost, ease of obtaining certification, waste management, water and energy management and length of certification. The result of this comparison is described in Table 9, showing some pertinent details of each certification.

Table 9: Comparison of certifications

	LEED	AQUA	BLUE SEAL
Prerequisites for enrolling in the certification	None	None	Only projects financed through Caixa Econômica Federal
Implementation/maintenance costs	Fee: U$D600.00 + value per m^2 : up to 5,000m^2 - U$D2,250.00; from 5,001 to oO.OOOm^2 - U$D0.45 per m^2 and from oO.OOlm^2 - U$D22,500.00	Variable amount: R$17,500.00 for built-up areas of up to 1,500m^2 and R$87,500.00 for built-up areas of 4,500m or more.[2]	Fee: R$40.00 + 7 x (n-1), where n= number of units built
Certification levels	Scores achieved at the different assessment levels. Basic Certification (40 to 49 points); Silver Certification (50 to 59	Different levels: Excellent: maximum performance; Superior: good sustainability practices;	Separated into six distinct categories, where certification can be obtained through three

	points); Gold certification (60 to 79 points); and Platinum Certification (80 to 110 points)	Good: minimum acceptable performance	criteria (mandatory or not)
Waste management	Collection and storage of recyclable materials, reuse of the building, management of construction waste and reuse of resources	Total elimination of waste generated after correct sorting, prioritising the purchase of products with environmental certification and properly disposing of hazardous waste (batteries, medicines)	It requires the selection of a specific site for the selective collection of waste generated and education projects for the management of CDW.
Water resource management	Efficient use of water, innovative treatment technologies.	Monitoring and controlling water use, rapid maintenance of hydraulic systems to avoid waste and the use of water-saving toilet bowls.	Requires individual meters, water-saving appliances (toilets) and permeable areas in buildings
Energy management	Encouraging the creation and generation of renewable energies, individual measurement and verification and the use of green energies	Using energy-efficient light bulbs, favouring natural lighting in the units, individual meters, limiting the use of air conditioning and opting for energy-efficient electrical appliances.	Requires energy-saving devices in common areas and individualised meters
Outdoor area	Redevelopment of environmentally contaminated sites, transport, reduction of disturbances caused by construction, management of bad weather situations, recovery and protection of open spaces, landscape and exterior design and reduction of direct light radiation output.	Preventive and constant maintenance of all external communal areas and non-obstruction of appliances (vents, thermostats, sensors)	Requires landscaping, leisure, social and sports areas
Social aspects	Encourages the continuous search for knowledge of new renewable energy sources and construction technologies	Training for users and managers of the project, as well as encouraging the pursuit of environmental education	Environmental education for employees and future residents
Duration or maintenance	Undetermined	Indefinite as long as it constantly meets certification requirements	Undetermined

(source: Author, 2016)

Table 9 shows some peculiarities where it can be seen that there are certifications that are more suitable for different types of buildings. Caixa's Casa Azul Seal is the only one that has a prerequisite for registration, and it is only possible to certify buildings that are financed by the bank, which could become an impasse if investors in this property wanted to make other certifications, such as LEED and AQUA, viable.

In terms of costs, however, the Casa Azul Seal is the easiest to apply for, since for large built-up areas (in the case of LEED more than 50,000m^2 and AQUA more than 4,500m2), the cost is too high to be viable. As for the criteria, LEED doesn't have fixed values, but is based on the sum of various scores at each of its assessment levels, which can lead to doubts as to its final score, while the AQUA and Casa Azul labels are based on numbers of criteria met or not.

When it comes to waste management, energy and water use, all the seals are similar in that they aim for economy, responsible consumption and the search for new sustainable sources and technologies for the consumption of these resources. As such, certifying bodies are encouraging the creation and cultivation of socio-environmental projects and actions with the aim of training society and investing in studies relating to sustainability in buildings of different types and uses. As far as external areas are concerned, all three programmes seek to improve not only the building itself, but its entire surroundings, creating dynamic, organised, safe and healthy communities for the region's residents.

3.2.5 Basic assessment criteria

Criteria and indicators can be understood, according to Silva (2006) as being a parameter or derived value that provides certain information about a given phenomenon and is developed to determine a specific objective. According to Cole (2002), in order to be useful, this indicator must allow an explanation of the reasons for changes in its values over time, be simple in the way it describes complex problems and use common compositions of key components and normalisation, thus allowing comparisons to be made. According to Mateus (2004), the use of sustainability indicators and parameters is based on definitions, rules, methods, classifications and the attribution of weights, both during their development and utilisation.

Silva (2006) describes that by providing the information and feedback needed to make the relevant decisions, indicators make it possible to: facilitate the setting of targets and the development of benchmarks for assessing and monitoring performance; measure or describe such performance; periodically monitor sustainability progress; provide communication with the client and other stakeholders; and, finally, derive direct benefits from sustainability and performance. Also according to Silva (2006), it should be noted that an indicator is not a number but a variable that can be added a value to measure items in terms of both quantity and quality.

The first indicators were developed in 1989 by the Organisation for Economic Co-operation and Development (OECD), which resulted in the publication of environmental indicators in 1991 (OECD, 2016). From this date onwards, as Silva (2006) mentions, the UN also began to figure in this scenario through the United Nations Commission on Sustainable Development (UN CSD), which emerged at UNCED in Rio de Janeiro. Silva (2006) states that in 1993, both organisations joined forces to publish their first set of indicators. This set was based on four different basic approaches: the media approach, which aims to organise environmental issues through the perspective of the main environmental components (air, soil, water, etc); the pressure-response model approach, which consists of the impacts of human activities on the environment (pressures) and their subsequent transformation (response); resource accounting with the aim of tracing flows from their extraction, through the stages of their processing, final use and return to the environment in the form of emissions and waste generated, or to the economy through recycling; and finally, ecological approaches, which include various models, monitoring techniques and ecological indices. In 1999, the Construction Related Sustainability Indicators (CRISP) project was launched, which provided for six types of indicators, as described in Table 10 below (Silva, 2006).

Table 10: Indicators set out in CRISP

TYPE	DESCRIPTION
Pressure	Describes the release of emissions and the use of resources and land
Performance	Product behaviour in its intended use
State	Informs about the quantity and quality of physical, biological, chemical and social, economic and cultural phenomena
Impact	Describes impacts caused by changes in the natural and built environments, impacts on biodiversity, availability of resources and provision of adequate health and safety conditions.
Reply	Response from social groups , endeavours and initiatives government to avoid or compensate for changes or adapt to them
Efficiency	Pressures on human activities, response or product performance, in terms of emissions and waste generated per unit of product

(source: Silva, 2006)

These indicators aimed for a global scale, aspects of environmental, economic and social development, as well as targeting categories of urban construction, infrastructure, buildings, construction products and processes.

A second initiative to create indicators, according to Silva (2006), came from the Construction Industry Research and Information Association (CIRIA), which carried out extensive consultancy in the UK for the construction sector, giving rise to ten key themes for building sustainable buildings. Subsequently, in 2003, the basic assessment items according to ISO CD 21931/03 emerged, which, according to Silva (2006), are shown in Table 11.

Table 11: ISO 21931/03 assessment items

CATEGORY	SUBCATEGORY
Internal environment	Thermal and acoustic comfort, lighting and air quality

Energy	Energy for operation, efficient operation, thermal load, use of natural energy and efficiency of systems
Resources and materials	Water consumption, productivity in the use of resources, avoiding the use of pollutants
Impacts on the surroundings	Pollution and loads on local infrastructure

(source: Silva, 2006)

Silva (2006) states that sustainability criteria should be chosen to express environmental, social and economic aspects, as well as describing the essential impacts of the building. This is used to assess which indicators are most relevant among those that fulfil the requirements of relevance, objectivity, accessibility, comprehensibility, measurability, sensitivity and traceability.

CHAPTER 4

METHODOLOGY

This chapter presents the methodology used to develop the sustainability evaluation criteria for various buildings.

4.1 Criteria adopted

The methodologies used to assess the sustainability of buildings can be used in any type of development, whether residential, commercial, hospital, school or even allotments, and can be used in any part of a building's useful life. However, Silva (2003) states that most methodologies are better suited to evaluating new projects or buildings, working with computer resources to assess this performance. For him, a complete cycle of indicator development falls into four activities: derivation and preliminary selection of indicators; selection or development of an analytical framework; implementation and validation of the proposed indicators through case studies, and; benchmarking of the values of the indicators created and their performance targets.

To create the indicators to be used for the general assessment of new buildings proposed in this academic work, we used the categories and criteria highlighted by Fossati (2008) and other environmental certifications as a basic model, which are shown in Table 12.

Using the requirements presented in Table 12 and the criteria described by the CRISP project (Table 10), a spreadsheet was drawn up containing the indicators selected in order to generate a methodology for assessing the sustainability of buildings. This selection was the most important part of the research work, as it aimed to choose criteria that could be applied to any type of building. Its importance also extended to defining criteria that were plausible for the standards and civil technologies currently used on building sites across the country.

Chart 12: Categories and basic criteria adopted

CATEGORY	CRITERION
Process management	Implementation of project and construction site improvement practices; Implementation of waste management practices; Water and energy use management; Building maintenance planning.
Environmental performance	Energy use over the life of the building; Water consumption and effluent management throughout its useful life; Losses recorded in the main services; Responsibility in the use of materials during

	construction; Emission of substances that cause the greenhouse effect.
Social performance	Satisfaction of the employees involved; Occupational health, safety and the workplace; Quality of the internal and external environment and services; Relations with suppliers, the local community, clients and end users.
Economic performance	Design/construction process; Added value and return on capital; Benefits resulting from investments in sustainability.
Commitment and productivity	Sustainability as a corporate priority; Contribution to building stable communities; Relations with society.

(source: Silva, 2003 adap. Fossati, 2008)

The choice of categories and criteria is based on Deponti's (2002) description of the criteria to be used, identifying the public that will be involved in evaluating and monitoring the system, geographically delimiting where the system will be used, determining the time scale (weekly, monthly, annual analysis), and finally characterising the production system (construction site), such as climate, vegetation, soil, etc.

Based on the models proposed by Fossati (2008), the methodology for determining, analysing and weighting the indicators for measuring the sustainability of buildings can be seen in Figure 7.

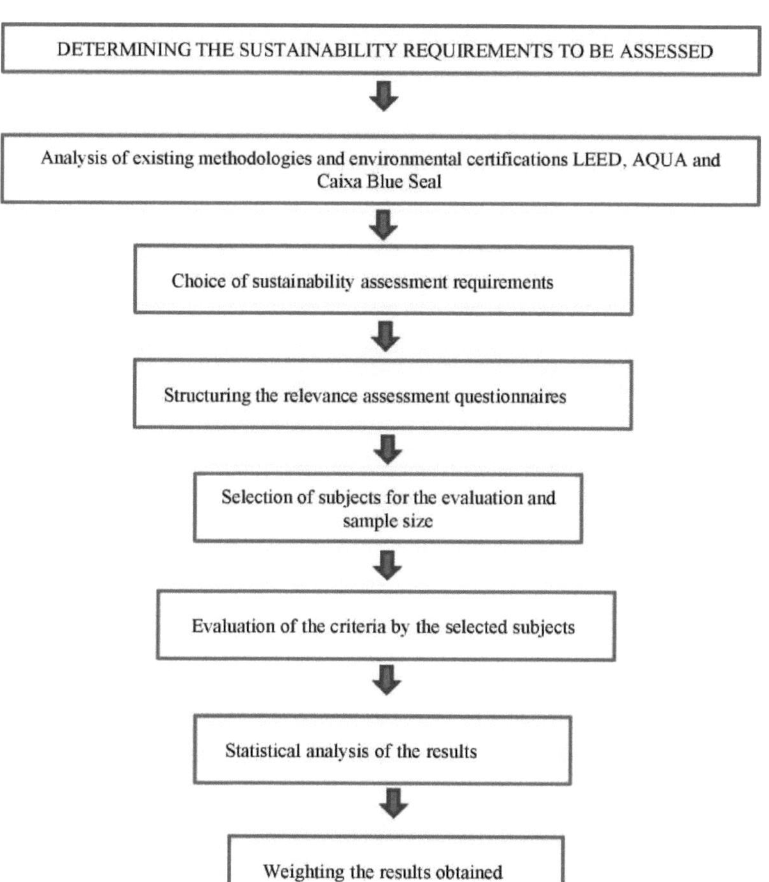

Figure 7: Flowchart of methodology steps

(source: adapted by the author from Fossati, 2008)

With regard to the first part of the flowchart, where the determination of the requirements to be assessed is mentioned, the steps of analysing existing methodologies were carried out; research in standards and legislation, and identifying the basic requirements to be used. These steps can be found in chapter 3 - literature review of this document. A comparison of certification methodologies in Brazil can be found in item 3.2.4 of the literature review. The next step was to draw up the checklist with the requirements chosen to create the assessment methodology. This checklist is detailed in Chart 13.

Chart 13: Evaluation criteria chosen

Category	Evaluation criteria	Numbering
External environment	Construction site with low environmental impact	1
	Improvements to the development's surroundings	2
	Landscaping	3
	External design of the building	4

Project	Projects with thermal and acoustic performance	5
	Leisure and social facilities in buildings	6
Water resource management	Individualised metering	7
	Installation of energy-saving devices	8
	Proper treatment of domestic waste	9
	Reusing rainwater	10
Energy resource management	Individualised metering	11
	Energy-saving devices in common areas	12
	Installation of solar energy generating devices	13
	Use of appliances in common areas with low energy consumption	14
Solid waste management	Recycling waste generated on site	15
	Selective collection	16
Materials used	Use of recyclable materials on construction sites	17
	Quality control of materials used	18
	Controlling material waste	19
Internal environment	Quality, safety and health on site	20
	Education of employees in waste management	21
	Professional and personal development of employees	22
Utilisation	Guidance and training for building managers	23
	Well-designed descriptive manuals	24
	Ongoing prevention and maintenance	25

(source: Author, 2016)

The criteria have been separated into eight categories that specify the area of each of the criteria, keeping the items better organised. The detailed description and examples of the application of each of the criteria are described in ANNEX 1.

The next phase of this work involved applying the evaluation questionnaire to the subjects to determine the relevance of each criterion. The main objective of this stage was to measure the relevance of each of the criteria indicated. To do this, different weights were adopted, as shown in Table 4. The greater the weight attributed to a criterion, the greater its overall relevance to the assessment of the building's sustainability.

Table 4: Relevance of evaluation criteria measured by assigning a weight scale

Relevance	**Not much = 1**
	Median = 3
	Large = 5

(source: Author, 2016)

4.2 . Sample definition

To define the sample, the principles of non-probability quota sampling were considered. **"A non-probability sample is obtained when access to information is not** so simple or resources are limited, so the researcher makes use of data that is more within **their reach, it is called convenience sampling" (GUIMARÃES, 2012, pg.** 19).

The sample size was selected based on a basic prerequisite among all the subjects, which was a minimum of three years' experience in the construction sector. According to Pandolfi (2015), when the population has certain known and homogeneous characteristics, a small sample size can be studied. Based on this, the number of strata needed to apply the questionnaire assessing the relevance of each of the criteria determined was calculated. The sample size calculation method was followed according to Pandolfi (2015) and is specified below.

a) Defining the number of classes: the classes defined for the study were divided into three distinct groups - Civil Engineering undergraduates, Civil Engineering and Architecture and Urbanism lecturers and professionals in the construction market with a degree in Civil Engineering or Architecture and Urbanism, as shown in Chart 14.

Table 14: Sample size calculation

Undergraduates	1 Class	Civil Engineering
Teachers	2 Classes	Civil Engineering
		Architecture and Urban Planning
Professionals	2 Classes	Civil Engineering
		Architecture and Urban Planning

(source: Author, 2016)

b) Calculation of strata: once the classes had been defined, the number of strata was calculated using the calculations shown in formula 1, where the numbers of all the classes are multiplied:

Number of strata = 2 x 2 x 1 = 4 strata (formula 01)

c) Calculation of the number of groupings: the variables with the largest number of classes are identified, these being the teacher and professional classes, with two classes each. The calculation is described below:

Maximum number of clusters = 2 x 2 = 4 clusters (formula 02)

d) Calculating the sample size: to obtain the size of the sample to be interviewed, Table 6 presented by Pandolfi (2015) was used, from which the average Coefficient of Variation (CV) was taken, as well as the minimum Relative Error Allowable (RE). The CV was set at 15.4 with an average RE of 5%.

Table 5: Usual CV and ER values

		High significance level α = 0.01; $Z_{\alpha/2}$ = 2.575			Moderate significance level α = 0.05; $Z_{\alpha/2}$ = 1.960			Low significance level α = 0.10; $Z_{\alpha/2}$ =1.645		
	ER	**Bass**	**Medium**	**High**	**Bass**	**Medium**	**High**	**Bass**	**Medium**	**High**
cv		2,5%	5%	10%	2,5%	5%	10%	2,5%	5%	10%
Bass	5%	**26,5**	**6.6**	1,7	15,4	**3,8**	1,0	10,8	**2,7**	0.7
Moderate	10%	106,1	**26.5**	6,6	61,5	15.4	**3,8**	43,3	10,8	**2.7**
High	20%	424,4	106,1	26.5	245,9	61,5	15,4	173,2	43,3	10.8

(source: Pandolfi, 2015)

e) Total number of questionnaires: using the selected CV of 15.4, the total number of questionnaires to be applied among the aforementioned classes is calculated:

Total questionnaires = CV x Number of clusters = 15.4 x 4 = 62 (rounded) (formula 03)

f) Distribution by strata: with the number of questionnaires to be applied (62 according to the calculation made), the distribution by strata was defined. This was based on the calculation model proportional to the number of strata, which ensures greater precision in the inferences applied, according to Pandolfi (2015). The formula used for the calculation is described below:

$$nqi = \frac{npi}{\Sigma npi} xTQ \qquad \text{(fórmula 04)}$$

Where: nqi = number of questionnaires to be applied in stratum i;

npi = number of the population belonging to stratum i;

TQ = total number of questionnaires to be administered.

The following distribution of questionnaires within strata and classes was calculated and defined, totalling 64 questionnaires:

a) Civil Engineering undergraduates = 16;

b) Civil Engineering lecturers = 12

c) teachers of the Architecture and Urbanism course = 12

d) professionals trained in Civil Engineering = 12

e) professionals trained in Architecture and Urbanism = 12

4.3 Statistical analysis of the data obtained

Finally, once the questionnaires had been answered, the results were statistically analysed in order to determine and weight the final values for applying the chosen criteria. Table 6 shows the averages and standard deviations of the answers to the 64 questionnaires applied to the sample. Using these values, it was possible to measure a final cut-off value for the application of the criteria in a real case of sustainability assessment of any type of building.

CHAPTER 5

RESULTS

From the data obtained in Table 6, it was possible to verify the validation of each of them, showing that their lowest average relevance was equal to 6.26 and the highest equal to 10. It was therefore clear that none of the criteria chosen could be replaced or excluded from the evaluation spreadsheet.

Table 6: Calculated mean and standard deviation

Criteria	Average	Standard deviation
1	3,781	1,525
2	3,188	1,158
3	2,938	1,223
4	3,219	1,586
5	4,469	0,883
6	2,875	1,452
7	4,063	1,273
8	4,406	1,308
9	4,406	1,259
10	4,594	0,947
11	4,344	1,1176
12	4,281	1,293
13	4,188	1,261
14	3,313	1,236
15	4,219	1,152
16	4,281	1,138
17	3,938	1,223
18	4,188	1,102
19	4,188	1,446
20	3,844	1,406
21	4,031	1,1172
22	3,656	1,372
23	3,781	1,397
24	3,938	1,273
25	4,000	1,275

(source: Author, 2016)

Using this data, the coded average values were obtained with a weight of 10 in order to define the cut-off point for the assessment, thus defining which criteria should be ticked in the assessment of the buildings. The coded average values are shown in Table 7, where they have been grouped from highest to lowest.

Table 7: Coded averages

Criteria	Value
10	10
5	9,73
8	9,59
9	9,59
11	9,46
12	9,32

16	9,32
15	9,18
13	9,12
18	9,12
19	9,12
7	8,84
21	8,78
25	8,71
17	8,57
24	8,57
20	8,37
1	8,23
23	8,23
22	7,96
14	7,21
4	7,01
2	6,94
3	6,39
6	6,26

(source: Author, 2016)

With this in mind, the cut-off point for the sustainability assessment was set at a minimum value of 8.5. This value was suggested by construction professionals and lecturers. Therefore, for a building to be classified as sustainable, it must fulfil all the criteria with a relevance of 8.50 and above. These criteria are described in Table 15.

Table 15: Mandatory evaluation criteria

Category	N°	Evaluation criteria
Project	5	Projects with thermal and acoustic performance
Water resource management	7	Individualised metering
	8	Installation of energy-saving devices
	9	Proper treatment of domestic waste
	10	Reusing rainwater
Energy resource management	11	Individualised metering
	12	Energy-saving devices in communal areas
	13	Installation of solar energy generating devices
Solid waste management	15	Recycling waste generated on site
	16	Selective collection
Materials used	17	Use of recyclable materials on construction sites
	18	Quality control of materials used
	19	Controlling material waste
Internal environment	21	Education of employees in waste management
Utilisation	24	Well-designed descriptive manuals
	25	Ongoing prevention and maintenance

(source: Author, 2016)

When these evaluation criteria are applied to real cases of projects aimed at sustainability, the buildings must comply with the mandatory criteria in order to be classified as sustainable. It can be seen that all the criteria relating to the social axis of the three pillars of sustainability were less relevant to the subjects questioned. This may be linked to the fact that these factors are less visible to construction professionals, which does not apply to the criteria in the categories of water, energy and waste management, which are more widely applied and evidenced on construction sites. The same can be seen in the criteria relating to the external environment category, which can be understood as a lack of planning for projects and their surroundings.

CHAPTER 6

CONCLUSIONS

In the course of this research, it was possible to analyse important concepts within the subject of building sustainability, such as its basic definition, the life cycle of buildings and the training of those involved in the whole process. We also studied three of the most widely used environmental certifications for the sustainability of various projects in the country: LEED, AQUA and Caixa's Blue House Seal. A comparison was made between these certifications, emphasising certain criteria such as implementation costs, different levels and classifications of certification, social aspects, duration and the main requirements taken into account by each certification.

In order to choose and create the evaluation criteria described in this work, we sought the basis of the three certifications studied, as well as the minimum criteria required to develop any sustainability evaluation criteria for various buildings. After determining these criteria and having them measured and validated by various professionals in the construction industry, as well as lecturers in civil and environmental engineering and architecture and urban planning, it was found that it is possible to develop a sustainability assessment system for various types of buildings.

In view of the above, this work has contributed to the development of important definitions and parameters on the subject of sustainability. It has also sought to encourage the planning and development of new projects with a view to the three pillars of sustainability: environmental, economic and social aspects. In this way, it is possible to make society aware of the damage caused to the environment in the construction of various property developments, but without having a major influence on the economy of construction companies or on society as a whole.

Finally, this work may help to evaluate new building projects in order to plan works that will bring various benefits to civil engineering entrepreneurs, those involved in the executive processes, the local community and, above all, the environment.

CHAPTER 7

REFERENCES

ALMEIDA, A. A; VIANA, F. G; PISANI, M. J. Social project criteria: A comparative analysis between the Blue House Seal and the social technical work model proposed by the Ministry of Cities. URBFAVELAS SEMINAR, 2014, São Bernardo do Campo. São Bernardo do Campo: National Seminar on Slum Urbanisation, 2014. P. 3 to 8.

BRAZIL, Ministry of the Environment. Sustainable construction. Available at: < http://www.mma.gov.br/cidades-sustentaveis/urbanismo- sustentavel/constru%C3%A7%C3%A3o-sustent%C3%A1vel> Accessed on: 27 April 2016.

BRUNDTLAN Commission. World Commission on Environment and Development: Our Common Future. University of Oxford. New York, 1987.

COLE, R. Sustainable Building: Indicators of progress, n. 4, p. 17, 2002.

CONAMA - NATIONAL ENVIRONMENT COUNCIL. Resolution no. 275 of 25 April 2001. Official Gazette of the Republic of Brazil.

DEPONTI, C. M; ECKERT, C; AZAMBUJA, J. L. B. Strategy for building indicators for assessing sustainability and monitoring systems. Specialisation thesis in Agricultural Engineering. Federal University of Rio Grande do Sul, Porto Alegre, 2002.

DIAS, R. Gestão ambiental - Responsabilidade social e sustentabilidade. 2ª ed. São Paulo: Editora Atlas, 2011.

FOSSATI M. Metodologia para avaliação da sustentabilidade de projetos de edifícios: o caso de escritórios em Florianópolis. 2008. PhD thesis in Civil Engineering - Federal University of Santa Catarina, Florianópolis, 2008.

VANZOLINI FOUNDATION, 2015. Available at: < http://vanzolini.org.br/aqua/referencias- e-guias/>. Accessed on: 18/06/16.

GREENBUILDINGCOUNCILBRASIL-GBCB , 2001. Available at <http://gbcbrasil.org.br>. Accessed on: 28 April 2016.

GREENBUILDINGCOUNCILBRASIL-GBCB , 2014. Available at <http://gbcbrasil.org.br/sobre-certificado.php>. Accessed on: 22 April 2016.

GOUVINHAS, R. P. Ferramentas da gestão ambiental competitividade e sustentabilidade. Natal, Ed. CEFET/RN, 2008.

GUIDE TO SUSTAINABILITY IN CONSTRUCTION. Chamber of the Construction Industry - CIC/FIEMG, Belo Horizonte, 2008.

CAIXA GUIDE - ENVIRONMENTAL SUSTAINABILITY (BLUE HOUSE). Caixa Económica Federal (CEF). Brasília, 2010.

GUIMARÃES, P. R. B. Quantitative statistical methods. Curitiba, IESDE, 2012.

INSTITUTE FOR THE DEVELOPMENT OF ECOLOGICAL HOUSING (IDHEA). What is sustainable construction? Available at: <http://www.idhea.com.br/pdf/entrevista.pdf>. Accessed on: 27 April 2016.

INTERNATIONAL ORGANISATION FOR STANDARDISATION, ISO. ISO 21931: Building and constructed assets - Sustainability in building - Framework for assessment of environmental performance of buildings. Geneva, 2003.

JOHN, V. M; SILVA, V. G; AGOPYAN, V. Agenda 21: A discussion proposal for the Brazilian construction industry. ANTAC - NATIONAL MEETING AND I LATIN AMERICAN MEETING ON SUSTAINABLE BUILDINGS AND COMMUNITIES, Canela, 2001. Canela, 2001. Not paginated.

LEITE, V. F. Environmental certification in civil construction - Leed and Aqua systems. 2011. Monograph for the Undergraduate Degree in Civil Engineering - Federal University of Minas Gerais, Belo Horizonte, 2011.

LEMOS, H. M. Socio-environmental responsibility. Rio de Janeiro: Editora FGV, 2013.

LUCAS, V. S. Sustainable construction - Assessment and certification system. 2011. Master's Degree in Civil Engineering - Faculty of Science and Technology, Universidade Nova de Lisboa, Lisbon, 2011.

MATEUS, R; BRAGANÇA, L. Evaluation of construction sustainability: Development of a methodology for evaluating the sustainability of construction solutions. 2004.

Organisation of Economic Co-operation and Development - OECD. Available at: < http://www.oecd.org/>. Accessed on: 29 July 2016.

PANDOLFI, C. Type of data collection - Sampling. Production Engineering course lecture notes. Serra Gaúcha College, Caxias do Sul, 2015.

Exame magazine; Brazilian GDP. Available at: < http://exame.abril.com.br/revista-exame/>. Accessed on: 11 August 2016.

SILVA, V. G; Sustainability indicators for buildings: state of the art and challenges for development in Brazil. Faculty of Civil Engineering, Architecture and Urbanism, Campinas, 2006.

SILVA, M.G. Life cycle analysis applied to the civil construction sector: review of the approach and current status. UNICAMP, Campinas, 2003.

SOARES, S. R; SOUZA, D. M; PEREIRA, S. W. Coletânea Habitare, Vol. 7 - Construção e Meio Ambiente. Porto Alegre, 2006. p. 102 and 105.

COLLECTIVE SUSTAINABILITY; Three pillars of sustainability. Available at: <http://sustentabilidadecoletiva.com.br/dimensao-ambiental/> Accessed on: 19 May 2016.

UCHOA, G; MACÊDO, L; BARTZ, C. The evaluation of sustainable construction in Brazil - Methods. In: XV ENCONTRO NACIONAL DE TECNOLOGIA DO AMBIENTE CONSTRUÍDO, Maceió, 2014. Not paginated.

CHAPTER 8

ANNEX 1

The items chosen for assessing the sustainability of buildings are described and exemplified below:

- Construction site with low environmental impact: this refers to encouraging the management of RCC, the organisation of living areas, safety on site and the control of material waste;
- Improvements to the development's surroundings: this deals with urban mobility, such as easy access for traffic to the building site, accessibility, improvements to the flow of vehicles along neighbouring roads, etc;
- Landscaping: encouraging the planting of small trees, green areas, reducing soil rectification, among other features;
- External design of the building: referring to the use of new architectural practices and technologies, giving the space a more refined, clean and sophisticated visual aspect;
- Projects with thermal and acoustic performance: deals with construction techniques aimed at these objectives, such as glass skins, masonry with high thermal and acoustic performance, double glazing in window frames, among other techniques;
- Leisure and social facilities in buildings: Spaces designed in common areas to improve the quality of life and social life of users, such as party rooms, multi-sports courts, gyms, children's areas, swimming pools, etc;
- Individualised metering: individual meters for each housing unit in the development, thus ensuring greater control of water use and reducing waste;
- Installation of water-saving devices: use of toilets with water-saving systems in all units, taps and other devices with timers in the common areas of the building, etc;
- Adequate treatment of domestic waste: installation of devices to help treat the sewage generated, such as a pit and filter system, grease traps, among others, so that waste is not discharged into the public network without initial treatment;

R Rainwater reuse: installation of rainwater collection devices, rainwater storage systems and systems for reuse in common areas, such as cleaning car park floors or irrigating green areas;

- Individualised metering for energy resources: same proposal as for water resources;

- Energy-saving devices in communal areas: use of energy-saving devices such as LED light bulbs, timers, external lighting with photocells, etc;
- Installation of solar energy-generating devices: encouraging the installation of solar energy panels in specific areas of the building to generate clean, renewable energy;
- Use of appliances in common areas with low energy consumption: opt for the purchase of appliances with the National Electricity Conservation Programme (PROCEL) energy-saving seal in the common areas of the development;
- Recycling waste generated on site: encouraging the recycling of class A and B solid waste[3] generated throughout the construction process;
- Selective collection: Selective collection of waste generated during use of the development by its users, with an appropriate place for storage until it is removed by the public bodies responsible in each city;
- Use of recyclable materials on site: use of recyclable materials on construction sites, such as gravel or sand from ground Class A waste, plywood sheets with recycled material, among others;
- Quality control of the materials used: technological control of the materials used on site through the minimum requirements of ISO 9001 and the Brazilian Habitat Quality and Productivity Programme - PBQP-H;
- Control of material waste: control practices to reduce the waste of resources and materials used on site by construction companies or those responsible for the building;
- Quality, safety and health on the construction site: use of good practices and specific programmes aimed at the quality of the services carried out and the safety and health of the employees involved throughout the executive process;
- Education in the management of construction waste by employees: encouraging the education and training of all employees involved in the execution of the work to correctly manage the construction waste generated in the processes;
- Professional and personal development of employees: social projects developed by construction or

[3] Class A waste: waste that can be reused or recycled as aggregates, such as from: paving and other infrastructure works, including soil from earthworks; buildings: ceramic components (bricks, tiles, blocks, cladding boards), mortar and concrete. Including the manufacturing and/or demolition process of pre-cast concrete parts (blocks, pipes, kerbs, etc.) produced on the building site; Class B waste: waste that can be recycled for other uses, such as plastics, paper/cardboard, metals, glass, wood, etc. Source: National Environment Council - CONAMA Res. No. 275, 2001.

development companies with the aim of professionalising their internal or third-party employees, as well as social programmes aimed at the families of these workers;

- Guidance and training for building managers: training courses for future building managers to improve their understanding of the building's functional systems and maintenance;
- Well-described manuals: well-described owners' and managers' manuals, with as much relevant information about the enterprise as possible to help solve future problems;
- Prevention and ongoing maintenance: documented preventive maintenance controls in the development for its longer useful life and the comfort and well-being of users.

Printed by Books on Demand GmbH, Norderstedt / Germany